The waterfall

Even the humble waterfall gives off gamma wave
radiation and produces helium and ozone gases!
Nature does nuclear fusion from water!

$H_2O + O_2 \rightarrow He + O_3 + E + \gamma$

There is no source of nuclear fission. No radioactive
Isotopes around! And the regular water
Or steam in turbulent flow. Nature power.

Biology gets 40% of its energy from this source!
Your own beating heart does
Molecular nuclear fusion.

21 February 2013

Dedication
Ralf Britton my last English teacher.
Professor Argent my engineering professor. Who taught me to think!
Dr. MatZinger of the NI H. Who engaged me in biology
Mike Scantlebury the leader of my creative writing group in Ordsall.

I have a bit of trouble getting this document proofread! As the only medics who understand it are American. I am relying on Chris SRN RMN for his medical input.
For too long physics has said that only it knows about nuclear fusion. But they ignore the production of nuclear radiation by biology. Or even ignore the nuclear fusion going on in nature-starting from water down to -20° C - not atomic hydrogen at 10,000,000° C.
The 10,000,000° C is the result of atomic nuclear fusion. It is not a prerequisite for nuclear fusion. Around the earth nature does nuclear fusion in the deep seas - producing nuclear radiation, helium gas and free radical oxygen.
The free radical oxygen ends up as ozone. Where ever we smell ozone nature is doing molecular nuclear fusion from water. We also see MNF from hydrocarbons and ammonia-in refineries.
Physics is a remarkably silent about the emission of gamma wave radiation, and the presence of 5.125 parts per million helium in the air. This helium is lost to space and replaced every day.
There is only 2.5 ppm carbon dioxide in the global air: as global photosynthesis in plants, bacteria and algae

metabolise this gas into carbohydrates-excrete ink the surplus helium and oxygen.

^{3}Helium gas? Yes. It is formed from two hydrogen atoms in water molecules. So photosynthe FFsis is a type of biological molecular nuclear fusion.

Animal mitochondria (which produced the energy for a cell) produce gamma wave radiation and helium. Again from water. So animal metabolism is another form of biological molecular nuclear fusion.

Your blood system does MNF, as the blood flows in a turbulent fashion around your body. This is vital to know -to get all the professors of biology studying molecular nuclear fusion. Power with no carbon dioxide!

Only a whiff of helium-that is lost to space: 25% of the oxygen you breathe in: and a trace of low power gamma wave radiation. Almost identical to that released as you boil a kettle of water.

Every water boiler in the world releases gamma wave radiation. And again, helium gas. There is no chemical source of either.

New Science please!

'Well here you are Jake!' Professor Sing even left you a coffee machine' Ian, the Chemical Engineering professor told him. Jake had been at the UMAT University for two years. In the postgraduate room/bunker. He could only dream then about his own copy machine.

He didn't even care it was an old 70s machine, which did not do Late's-or even take your cloths for dry cleaning! Just as long as it made coffee-potentially.

By the look of it, Rangive have only left it as it was broken! The postgraduates have been joking about the professor of the broken coffee bar for years!

'why the sudden promotion? Asked a slightly puzzled Jake. Presumably Rangive have received a better offer. Apparently from the Deli University. Where he would be near his own family, and by 2020 Indian was set to overtake China in GDP. America would be headed down to was the third world.

Hopefully passing England as it is had a higher GGP than Germany in 2010. Hopefully by 2014. Though England did not have the extra expense of the Euro, falling so close on behind the costs of reunification. Which would rumbled on to 2050. At conservative estimate. Would that Germany recover? Well, there recovered after losing two world wars.

It would have paid off the debts of the Euro by 2015. The lower development level of Eastern Germany realistically would go on for 200 years. Anchoring the mass of vibrant economy in the world two massively

cheap currency, was really going to mess with world economics.
'UMAT was there at the dawn of the industrial revolution an inability of Manchester; which along with Birmingham gave the world technology! There is were the days.
'that's why I came here' Jake added. Ian scowled.
'that were as in the 19th century! UMAT last did world shaking scientific research in the development of the expansive steam cycle' the professor lectured.
Repeating the text he gave to the undergraduates in the first year. Only now it had an unexpectedly a hollow ring to it.
'unless we get new world changing technology, UMAT will just be totally swallowed and digested by Manchester City university. That was never the idea! We were going to give Manchester something special. Access to technology they did not have' thundered Ian. Jake sat down and try not to look too scared! 'I thought global warming was the big thing in technology! Interjected the new lecture.
Ian laughed. 'The meteorologists are sure that the world stop warming in 1997' he explained patiently, like to some brain dead fission fan/moron: after all, Jake have been an undergraduate within the decade.
'but climate change' Jake protested.' 'Global warming in a cooling world. More nuclear invective about carbon dioxide. The world' never believe the academics about global warming, even less so now! Ian conceded morosely.
In truth, manmade climate change conceded that manmade global warming was wrong. The science

about global warming have been so strong-but totally illusory! Nuclear power have got the blame on the carbon dioxide again-what was there a problem with the gas of life?

It couldn't warm the earth between 1986 and 1998. Then produce 13 years of global cooling. Science does not change to affect the finances of nuclear power. Science is truth. So neither global warming or climate change could ever have been silence.

'we need new real science-that will shake the world!' The professor almost shouted. 'because otherwise' he gesticulated with his thumb crossing his throat. That is why the office is free, the future wires to China and India'. 'not even America' as Jake. Who had shares in American drug companies.

'yesterday's men! England have the world in the 19th century. And we let it go' sadly summarised Ian. 'so you are going to give us back our edge' hoped Ian. 'I will do my best sir' Jake answered the professor like you was a schoolteacher.

'So you want perpetual motion or nuclear fusion from me?' Checked Jake. 'perpetual motion?' No work has been done on that says 18th century engravers had a go at it. After nuclear fusion: the great hadron collider has turned out to be the worst hundred billion dollars ever invested in science.

'they couldn't even find the Boson particles! Jake brightened-he thought they had found the boson mason. 'the image years they found well for a particle that was half the size required. Physical fantasy time' Ian finished.

At least Jake had his own (broken) coffee machine. He locked in the drawers of the desk, He was pleased to see three unfinished crossword books. And yes he is going to save UMAT, and brush up on use words scills.

The forum

Dr Vince Turner strode into the room. He scanned the audience. Looking for ecological activists, or anyone be who have not been invited. They are not here. He nodded.
f'All clear' said Stephen, who has scanned the room for listening devices and cameras. Paying especial interest in any potential Internet links. There were none. Why would that be? This after all is a restaurant for the cleaning brigade. So they only had 3 hours to vacate the premises.
'Hi everybody that's not talk about fusion' they must started. Dianne objective 'I fault we were here to talk about that subject!' She pointed out.
'please forgive her everybody, she is new to this group! And she is yet got to learn what we are about. The X Group exist surely everybody, to be invisible!' He started to unexplain carefully.
'while the world is convinced carbon dioxide is the biggest threat to man. On earth, nuclear power gives us a blank cheque' he gloated. Dianne objected. After all she did supervise university modules into biology. When her hectic crossword schedule allowed.
'surely everybody realises that photosynthesis metabolises all available carbon dioxide' she objected. After all this was a university, and they were about truth.

'if everybody had stayed awake during O level biology they would realise it! Locally everyone he had leafed through to the pages on sexual reproduction' he explained with that for a way look of disgusted interest on its face.
'surely a biologist will inform the general population that the limit to life on earth is...' die and started only to be cut off by a dismissive hand wave. 'since 1986 all physics, chemistry and engineering departments have only been kept economically afloat due to the fictitious work on the climate' he patiently explained! Not very patiently.
'so burning carbon fuels is the greatest threat to life on earth' he finished. 'but biology teaches us that the circulating organic carbon expands life on earth!' Dianne objective.
'don't get two hunger and truth! We are talking research budgets here. Now on to nuclear fusion-the X thing!
'now this can only be achieved at tens of millions of degrees C and turns of atmospheres pressure! He recited mindlessly.
'I have raised some work this week or steam engines producing helium gas engine off gamma wave radiation' objective Dr. Leonard Pishcow. A look of abject horror and repulsion crossed Vincent's, let's call it a face.
'does that get research money' the objective. 'well yes, for a seminal PH D!' Leonard promptly declared. 'and have then what?' Vincent asked him.
'I will tell you' he carried on, not even giving them a had time to take in a breath of air. Or whatever he used to breathe with. 'we will all be out of a job! Is a scientist a physicist? Or just stupid?'

'a chemical engineer' stammered Leonard. A look of abject supremacy would have flashed over Vincent's expression, except for that recent accident with a liquid nitrogen and anaesthetic.
'applying pressure to divert the 'scientist' into a more useful research area. The nuclear slush fund has $10,000,000 are spent in it. You're as some of it to diverert him' Vincent finished.
Stephen Watchman objective. 'Isn't this diverting useful research in be counterproductive?' 'do you want a fabulously over remunerated retirement? Or don't you?' Vincent replied.
'science today he is about money. We drip feed two to the population according to our time schedule. The discovery of nuclear fusion is pencilled in for 200 years away! By which time the failed fusion torouses will circle the earth three times, and have cost the world four trillion dollars' Vincent gladly recited.
'until then, this group exists to manage the gradual release of the science to the great unwashed. Here fantastically over pay us' he noted O so truthfully!
'people who somehow imagine carbon dioxide-the gas of life could cause any trouble had it coming to them'
'the levels of that gas in the air are static at two parts per million, says levels fell after the little ice age' die and proudly told the room. 'if you want a career in academia, you would do well to forget that fact. Otherwise how could you write fictitious papers on the climate without laughing' pointed out Vincent.

'if there is no other affairs we need to deal with, mind is a kind of Heineken' Vincent declared drawing the meeting to an end!

Vincent would have strode purposefully to the cafeteria, and maybe there was no great power does he have in mind will ever imbibing for fluid. And avoiding doing his undergraduate essay marking by another couple of hours.

He was always on the case of undergraduates to get the essays in on the appointed deadline. Really it didn't matter! He got proficient over the years at the incentive arising students to turn in plagiarised essays about something or other.

The advent of the Internet has really improved the date pool for plagiaration. You could have now by essays over the Internet! Which had ended the income stream for graduates to sell their old course essays to the new undergraduates. That was significant progress!

It didn't help graduates to pay off their student loans. He smiled wistfully for an instant, remember the halcyon days of full grants. This had allowed him to escape his family history! His family had ended to be accountants. If only he had realized then that the best paid careers were as plumbers or electricians. That only required a 2 year non vocational qualification-and no blasted essays to mark!

He was hoping for some diversion! A major world famine, hurricane all drought just to relieve the monotony of academic life. He handed the final refrain from Beethoven's fourth symphony-he was at stressed.

'Vincent!' Said Prof Ian Taylor-the head of Chemical Engineering. Very good if you were ever in a hurry for non contact explosive or ecological pantheon! Best not to get the two mixed up though!
'I would like you to meet J theake' he told his old friend! The guy was a physicist-but nobody is perfect. Not the gallows to change the world, but after all engineering was practical physics. He would have humoured the guy, but there were in the same golf stream. Now that was world changing only important.
'Hi Jake' Vincent announced as he warmly shook hand. These guys did tend to come and go. They came to UMAT university to change the world, and progressively drowned under marking. And left forearm over remunerated career in advertising, or television or whatever!
UMAT had an illustrious past-before it was amalgamated with Manchester Civic University. Some of the 19th century science which are of the industrial revolution having done there. UMAT was the title they have adopted, and it was not until 2008 they thought a better name-but by then they became part of MCU.
'Jake has come here as a postgraduate researcher to change the world' he announced with no perceptible trace of sarcasm. His eyebrows said it all. Vincent nodded wearily. Another one for the academic mill!
'I have high hopes are doing important work at this university' Jake declared. 'so you are the next James are you?' Vincent checked. Jake looked surprised.

'the last person who did important work at UMAT. In fact, the only guy who did important work before me! He boasted. Jake opened his mouth to query this.
'I and just about to publish a world important document on high a temperature semiconductors" Vincent declared. 'so you got results did you?' Ian checked. Knowing full well from the heated words from his post graduate researcher, that he had not.
'Brownian motion killed us. We were getting good results, until we hit 250° C. Then the electronic noise from the atomic vibrations within the semiconductor destroyed our power again. We were down to 1.2 times, went away for the weekend and lower at sea getting power reduction rather an amplification.' Vincent conceded in an unusual moment of academic frankness.
'have you thought of doing any more work on the liquid clutch?' Ian after him. 'after all, that did make it on to tomorrow's world in '98-so very recent' he finished.
'we couldn't get over the powered demand problem' Vincent conceded. 'did you think of engaging a hydraulic link, to join in the gear box and drive shaft together-so you could then this engaged the clutch power line' Ian proferrred.
'interesting' Vincent noted. 'there is research, funded research, yet to be done there. Now Joke-what are you doing?' 'well I was thinking of tea drinking' Jake replied as he ordered 3 cups of tea.
'I am investigating the production of helium gas and release of nuclear radiation by nature' Jake proudly told him.

'Oh, so nothing important than' Vincent observed as he got down to last Tuesday's Telegraph crossword. He only had three words yet to fill in. An eight letter words mean many skilled -POLY****'. He signed. Time for a quick phone call to English!
After all that sort of words do not apply to any body at UMAT in last 200 years. Enough thinking!, But there were. Activity for academics after doing crosswords!
The Vincent was not impressed by climate change. Nuclear power adopted it was the world stopped warming 1997! 2010 100 leading American academics point yet the world had been cooling since 1998. 2012 even the UK met office conceded natural global warming has come to the expected end.
Man made climate change was based around an increase of carbon dioxide in the air. Even a professor of physics realise that biology prevented this! There have been a static trace of carbon dioxide in the air for last two centuries.
Climate change appeared to me in the global wetting in Europe. And global drying in Texas. In other words the natural weather! Not global anything.
Man kind has no control over the trace of carbon dioxide left in the air. Then send were scratching around for the new research area, as there scientist have research global warming since 2003. IA day it is very many had even bothered reading about climate change-as he predicted nothing!
Climate change was based around the erroneous notion that man could affect the trace of carbon dioxide in the

air. This and had learnt as an undergraduate that the fossil fuels has formed at the end of the Jurassic age. Physicists were interested because all the fossils were from life that died in late Jurassic. He knew from his a level history that the Jurassic age had had three natural ice ages. Sea levels 60 metres below what they have today.

You've so nice work on there being 65% more life on earth in the Jurassic even on my master day. Climate change more spurious fiction on behalf of nuclear power. They most toxic industry-but physicists had unwittingly helped develop!

At least the Times crossword was still straightforward! And true. Climate change said nothing, was obviously wrong from its inception! Global warming in a naturally cooling world.

Then send wandered along the courtyard path having the third chorus from Raviles anlum be in D major. The flatulence overture! He always pundits when he was annoyed. He actually got it wrong, but when confronted with this fact he explained that he'd got it right. The composer had transcribed it wrongly!

The wine were scientists not allow to study true stuff? Or just given huge amounts of money for doing very little-as you used to be the case in the days of full grants from the government. Tony Bling had got that one right! Today income scientists have to study whatever they get funding to study! No matter how preposterous an impossible and it actually was.

Academics now have this agreement: don't comment on somebody else's page research. Usually because the person to prove it was demonstrable garbage.
Global warming and climate change were such absolute rubbish venting couldn't believe they got 36 years research out of this. And 4 world conferences.
This had at least allow them to get some serious breach practice in. An academic life was a hard life! Occasionally you were even expected to do real, true and useful work. That was not the academic way!

Drinking

'Jake meet Susan. Some other postgraduates said you might be able to help her with her final dissertation' Vincent told Jake as he delivered 2 pints of breast and the girl with an alternative fashion sense.
This may be and was of it hard on Susan, and students are notorious for dressing in whatever they can get at Oxfam. This year fashion from the 1970s is very popular! And we all remember have power that warms.
'I presume you are in Earth Sciences;-that's why we haven't crossed pass before!' Susan reasoned. Jake shook his head. He was missing a night of combat robot 'Death Combat' for this! She smiled at him, and for a fleeting instant it was worth it!
'hardly' Jake replied. He had learned in academia. Never give a straight answer to what appears to be a straight question. But it was after the working day, he was in a social situation, drinking real ale. This was the ideal instant to get a Vegan postgraduate student to agreeing to do trials for the Meat Marketing board.

'I am doing research into Chemical Engineering. I have found that nature does nuclear fusion from water' Jake told the thirty-something girl what he was actually about today! He hoped his reply was intriguing enough to get this help of this-he looked carefully; girl!

'I thought physics could only envisage nuclear fusion in suns at 10million° C. There was no way nuclear fusion who have an outside a laboratory' the girl recited established theories as if there were fact. It was far easier and thinking for herself.

'every 3 minutes somewhere around the world we have a lightning strike'. Jake could confirm this from his radio days, and that extremely low frequencies for lightening strikes -their radio emissions were there to be heard.

'I am a biology Ph.D. student, why does that matter? Susan challenged him. Thinking Vincent had got her chatting to somebody with no use to a PHD student investigating deep sea fish. Or maybe after all, she had had missed the high incidence of lightening going on in the seas.

She did not just walk away, as she realised deep sea fish in the arctic seas give off nuclear radiation! There being no radioactive isotopes around, this has intrigued her curiosity. Enough to remember that she had no idea about why this happened.

'all life on earth evolved in a period of great lightning activity.' Jake instructed her -him feeling his research work was being belittled. He was failing to see the relevance either. At least he got a drink of real ale. No if she compiled across as as a hobby his night was complete!

'as steam plasmas fire through the air, and the nearly inert nitrogen reacts with the oxygen gas in the air. To form NO and NO2' nuclear radiation and heat; he recited the greatest insight ever in scientific history! Maybe if that was a bit extravagant! But certainly it was her biggest inside he had ever had.

'nitrogen oxide -that is laughing gas! I always wondered why lightning made everybody feel happy Susan nearly conceded. You did get loads of lightning strikes on to the see, so there is probably explained why fish look so happy. or was her brain being sarcastic here! Obviously the real ale was having some effect. All that nitrous oxide ran down the rovers into the C fertilising sea life. This was an area that have fossils for four years. Or maybe not years, hours-now says she is starting this drink actually.

'this is going on during a period of heavy rain' Jake added. Susan looked interested in a vacant sort on the other side of the bar. He decided he had better explain. Jake realised he was losing her attention, and a better act quickly to engage her interests'.

'as rain fall through the air, there is loads of turbulence' he uttered his magic words. Realizing the significance eluded her. If he has said biology in the offering free pole dancing lessons for men she might have been intrigued! The affect of the real ale must have been waxing.

'the chaotic interaction' still nothing! He might as well have been talking Greek. ' the water drops collide and some do so heavily they form helium and free radical

oxygen-O' he detailed. Her hand went up subconsciously.
After all she was a biologist. And she can seem that rain science might be important, but not to D. C. arctic fish! Working to reduce European subsidies are fish and chip shops would have been more engaging just then.
'I thought there was no source of helium outside the summons or nuclear reactions' Susan protested. She had stayed awake for some reason in the physics lessons at school. She had tried to be interested in boys or pop music to avoid her lack of interest in physics showing through her jumper.
'there is 5.125 parts per million' he recited, now in science he knew. Susan's interests lights started to fade from her eyes. '2.5 times the level of CO_2-the global average level of carbon dioxide' he explained, and was rewarded to see the interest light up again!
He took it, drink from his real ale. This girl was a challenge, as there, and scientific interactions were so limited.
'I can see that is interesting! She conceded 'all biology and is based around photosynthesis metabolising carbon dioxide. Mankind could not increase the level of carbon dioxide in the air' tube filled in, as if only an idiot can believe otherwise.
'but helium has no use it does not react!' She pondered. 'I guess it just builds up in the air' she pondered the question-never had the bravery to ask physics. She vaguely wondered why the air had not filled up with helium all through the ages.

'helium does not react with anything' Jake conceded 'but earth's feeble gravity does not contain it! And it is lost to space the same day that it is made!' He informed her. She took a large sip from her ale, and settled down to listen. He was saying that the earth don't interstellar hydrogen into helium! Neither of which stood around in the earth air.
'each day the helium is made by doing nuclear fusion on water!' He laid out his of the science insight. Not knowing that she is a keen fell walker, and he would have taken an air sample on the heels to confirm all this is she had wished. The idea had never occurred to her. Obviously he was either a free thinker or -No! He was a free thinker.
'so this is a source of all this single atom oxygen, that makes all the ozone we see through nature' she checked-trying to utilised the little that she knew about chemistry.
'exactly' he was so happy. 'all the ozone we see through nature is a biproduct of the nuclear fusion going on in nature' he finished. 'Hmm' she intrigued 'Go on!' It was all so obvious, once it is explained in detail to you. Real scientists must have known this for 30 years! But still allowed nuclear power to give them all seen amounts of money to research fictitious stuff about the climate.
'all the nitrous and nitric acids fertilises the ground and the seas, without which there would be no life on earth' Jake told her. Trying to demonstrate in the same way, that he knew a little biology. Important word here 'little'.
'so you are trying to tell me that lightening creates all life on earth-that in turn is set up by heavy rain fall' she tried

to summarise his thinking. She could detect about 10 that good science papers waiting to be written here. 'molecular nuclear fusion goes on all through biology, and in the physical world: at breaking waves, waterfalls heavy rain and the resulting lightening' he finished.
Swig his beer again! Not the best brew!' There was two great 'a hint of rubber boot' for this pint to a troubled CAMRA (campaign for real ale). There is always such a simple world'! Water valley and yeast and the world give you money.
'Ah! That is why the Vincent introduced us! If all biology does your nuclear fusion from compounds of hydrogen, deep sea fish would be liberating massive amounts of heat and nuclear radiation' she half under stored. So the finish, like all animals, where nuclear!
'what is the title of your PH D?' Jake actually bothered to ask the girl after their 10 minutes' conversation.
'the fish ecosystem and nuclear processes beneath the North pole' she informed him. 'yes biology has known for 50 years that fish give off nuclear radiation' he conceded.
'just nobody could figure out why' she added. Thinking it was time for more ale and a chat with a Ph.D. student who is obviously not so focused on nuclear fusion. Her brain was buzzing, and she needed no more nuclear fusion science for a week!
She had always rather figured that she he was an interesting topic of conversation. For more horny undergraduate boys in the BeBop Mental coffee shop anyway.

She scanned the room, and glimpsed one such example of humanity toasting his backside on the fire. She though he was a serious masochist, or he had sat on something damp!
It had come to that unavoidable moment in academic life. That four letter word: work! He signed and decided he should go and check on whatever it was that he should be doing!
Academia smelt of dust, sweat and toil! Always collective minds trying to stop the biggest problems out there-and to the times crossword in record time. At the same time bring up their families and showing them the proper care low no attention they need to thrive.
They were like some implies in of biology case. Only more demanding and expensive. A diet of non stop shit was only an acceptable if your wife was a terrible cook! Even then children expected only the best faeces. Luckily the university has unless supplies of that!
It is one of the tragedies of academia, the people who pay your wages occasionally expect you to do some significant work. That can really get in the way of practicing for the next bridge tournament. Or during those other important things we usually call life. University work deadlines can interrupt a television mini series at an important juncture. Workers of the world would relate to this, but not understand the high conference and side interest commitments academics have to live with.

The laboratory

He boiling a flask of water-just like making a cup of T with no tea bag. It reminded him of his days on low caffeine diet. Before somebody told him that T does not actually contain caffeine. But without tea making his routine, there were not enough lecture commitments to fill his day.

He must have done this hundreds of times. It is what chemists are supposed to do! But today his interest was actually in the water boiling. If there was an Olympic metal what a strange sporting event the all and it would be.

The fact that he was usually using the boiled water to actually make a brew was an incidental side benefit. All that was what he intended to tell any incidental observers. All these years he have been boiling water he and never thought to direct a Geiger counter at the water as it boils. No he did do!

Jake shook his head. As the water came to the boil, the Geiger counter went crazy. Much as is brain was feeling now. The laboratory manager popped her head into the laboratory, and he in surprisine looked up. Couldn't Tintan tell he is doing groundbreaking work here?

'time to go home! Don't you have a warm meal for a wife to go home to?' She barked out as she turned the tumblers in the filing cabinet locks. You had to protect all the trade secrets against the unscrupulous cleaning crew.

She wasn't sure this is actually true! But it was what she is paid to ensure. Of all those Times crosswords must have some value! She occasionally sent a completed one in. And the only won the prize twice!

A wife? 'why are you are offering?' Jake checked. After all a warm wife might be an asset at the Christmas party.

His last girlfriend had gone off to Africa and would not be back for three months. Unless if she caught malaria or a interesting disease.

'ah you wouldn't want me' the Tin lady retorted; as she shut the divider and locked the doors. 'I taste horrid' as her last boyfriend had told her. She didn't know why, but postgraduate researches have a high incidence of their partners spontaneously combusting or cooling down with terminal boredom.

Jake shut down up brain and went to the senior common room. Then he picked up his rucksack and headed for the hills. Enough thinking! Time for some serious walking and even more serious drinking. In the hilltop public house.

He pulled up at the hotel in Eadale, and checked into his usual room. 'no more nasty science-enjoy the hills and the rest' Lynder instructed him. The youth hostel acted as a refuge for for all the brain addled researchers from UMAT. It was proud of its lack of intellectual stimulation. The poor students got enough of their mental stimulation exercises during the week. At the weekend all they wanted to do was walk, eat and drink. Some tried to combine all three activities at once an had to be rescued by a hikers emergency crew.

They would learn never to be so stupid and so overconfident again. Or just become university lectures! It is a basic entry requirement for teaching in tertiary education: A total lack of common sense!

The next morning he wandered down to the youth hostel lounge area. To his usual breakfast, bacon, sausage, egg and Haggis: A Scottish as usual! MacIntyre had owned a hotel in Scotland, before coming down to be near his grandchildren. He had never got into the habit are serving a full English.

'black pudding and sausage? Ground up in testiness with dried blood' he would declare has high volume. All the members of a hikers will begin salivating at the very mention of those words!' MacIntyre would mutter dark things about the Southenarks as he fed them every day. It was lucky for Jake that he did not come from Leeds- where tripe and onions were the breakfast specialities. 'the full Leeds' turns to induce rhythmic nausea in people not used to Yorkshire ways.

The or just people with functioning gastro intestinal tracts! That resented being fed on left over care stomachs and rancid vegetables.

'are you off to the Hills today?' The landlord enquired. Jake pondered for a moment, and use it to be humourous. 'I was going shark fishing in the river' he told the preoccupied man. Who have developed a selective deafness to deal with his guest utterances. It was not as if he was about to miss anything important- was it!

'very good! Be back before 4pm, because the weather it is going to turn Scottish on you!' He was warned.

Taking care to place a packed lunch I his rucksack and he was away. Not even having time to enquire what a Scottish surprise was. No doubt his stomach would find out in the course of the day. Annie had the local hospital A&E on speed dial on his mobile phone.

He got halfway up to the plateau near Edale and he left the tourist trail, in search of adventure. He ran into cow

Tony, one of his hiking friends! So called because he kept sheep. No he wasn't called Anthony! He made a mental note to enquire further just as soon as he was interested.

'Be back before 4' Jake was reminded. 'the weather is going to turn all Welsh on us' Tony instructed him. Jacques pondered about the geographical shift of the weather in the last 5 minutes. He mused if you ran into French Hellena the weather would no doubt be typically Parisian.

He scrambled up the track and sat down on the clearing by waterfalls! The waterfalls around Yorkshire rearley run dry, as all the politicians supply them with ample hot air, to condense into rain. Quite marvellous. If they could just attach a turbine to their utterances they would have perpetual power sussed!

He took out his lunch and I was startled to see the Geiger counter there! He had meant to leave it behind. He actually flicked it on. The needle jumped. As did Jake. There are no radioactive samples around here. He knew he was a fairly dangerous capture, but he doubted that even E would activate a Geiger counter. Not at the weekend anyway.

Maybe the rocks were radioactive-after all the natural geography made Aberdeen more radioactive then the seascale nuclear plant. Though obviously not during while regular radiation leaks. So not mandate until Friday! At the weekend.

He directed the Geiger counter towards the rocks. The reading died away. He sold it back around to locate the radioactive isotope: such elements are quite valuable.

But it was the waterfall itself which is giving out the radiation. Not some radioactive rocks. Unless you could find a way to package the waterfall, it was the method of doing nuclear fusion that was important! Putting water through pressure cycles! He began to see the idea. If you collided water molecules hydrogen ends fast into he each other, as the false above the strong atomic force that could hydrogen atoms supports, hydrogen is water molecules of water which fuse into helium.

You would also liberate atoms of oxygen gas! They would bond with oxygen molecules from the air, to form ozone. So where ever we smell ozone he nature it was doing molecular nuclear fusion. Doing the nuclear fusion of hydrogen atoms from compounds of hydrogen. As mind store bac innovate and in his skull: nature was fall of ozone! Linked to the production of heat. And he would put money on they also being the liberation of nuclear radiation.

Helium atoms were not held by the earth's feeble gravity: his high school physics had told him that much. We gain hydrogen from space. And at lightning strikes forms water!

And molecular nuclear fusion turning water into helium and oxygen. And so it all went around again! A heat generation system that produce no carbon dioxide, and no toxic fission waste. It utilised no over expensive carbon fuels.

Water fell regularly from the air, for free! The base substance to do molecular nuclear fusion. For free.

With no toxic waste. Not even the carbon dioxide which stimulated plant growth!

He pointed the Geiger counter briefly on waterfall, just so he could hear the click click click all the extra ecological powered jobs,. He was a scientist, he had to examine this. Formulate the equations. Write the scientific papers. Receive at least three Nobel prizes. He walked over to a waterfall and turned the Geiger counter on full blast again. They needed to hear about this Scientific Medical!

There was no doubt, the waterfall was giving off radiation. Not the surrounding rocks! The waterfall. He went back over to his rucksack and called out air sampling device, he had are there for the Wednesday practicals. The analysis would show the presence of helium and free radical oxygen.

He could almost hear the income trying I his first Scientific Paper. OK, he would write it on computer. Though it knew of no such evocative modern analogy He took an air sample. He stowed the jar back his rucksack and along with the serious stuff-walking and relaxing.

He had come for relaxing walk in the hills. An estate stumbled across a greatest scientific discovery ever! And he also got to enjoy a pint and a Scottish lunch meal. All without vomiting! That was the definition of a good day.

Smiling he descended back to the railway station. The train was half an hour late. So he sat down and work hard are not smiling too much. Curie, Darwin and Pitiful

the scienc names that they live on for all time! And the greatest of those words Jake Pitiful.
Not for the first time, he thought about changing his surname by deed poll! All he had to do now was two is a suitable name. 'Flash', 'Bolt'-no 'lightning'!
That was the name! Jake Lightning. The flash in the water fall. There fusion man. The greatest genius the world have ever seen. Though obviously one who needed new boots. His present pair were leaking.

Gas analysis

He stowed the air and sample away until he could do something with it! Other than blowing up a balloon for his nephew's party. He thought it likely the sample will contain helium and oxygen. But he could be wrong. Despite his mental boast he was well aware that he is not a genius. Yet.
He was aware that he was dipping his toe in the lake of 'as yet unknown science'. It all seemed so simple! Here he views his notes of the field trip from a waterfall. It up in a lovely summer day! Only a spoiled by the regional of occasionally have to do some work!
After all, that did fund his 20 hours a week unabashed. Maybe one day he'll an endless summer day drinking beer by a waterfall might earn him a Nobel. The he had once met a second cousin three times removed from a Nobel prize winner.
Who got his award by approving a medieval Latin manuscript was a fake: there not being too many functioning microwave ovens around the Roman forum

in July: any idiot knowing that name for a month was only invented in the 16th century.
He chose to fill his nephew's balloons with helium gas. To give everyone a laugh he filled every fourth balloon with helium! Guaranteed to make the balloons rise to the ceiling, but the fun would start as the balloons were popped.
It worked best if the annual karaoke session was in full swing by then. He did A Robin Give impersonation to die for. Which he very nearly did most years due to adhesion. He is biochemistry not functioning to whirl of inept helium.
Inflating though balloons with the laboratory helium side, he marked the helium balloons with a black cross, so he could be sure to pop one over Sammy's head just before the cake cutting. This should put him in a good mood! He remembered at age 6 that his own father, had squirted some gas from a cylinder into the air, and given him a really great birthday. Though his pet rabbit did suffer convulsions for the next three months. Nothing about science is too safe!
He mused there was nothing new under the sun! Like father like son. He just hoped he would not end up having to flee the country after an embezzlement scare. Like is dearly vanished stepfather.
How he had afforded all those teenage prostitutes in Bangkok, he never did figure out. Never be able to talk his father again before the excessive flatulence on volcanic eruptions scare. Which a killed his father and two semi naked teenage girls.

Next Monday in a moment of academic irrelevance, he fed the air sample into the spectrum analyser. It came out 70% helium, 21% oxygen, and 1% helium. He smiled with satisfaction! His hunch had been exactly on the money.

It was also good to know that the laboratory equipment did work! Though it had not been serviced since August 1952. There always best to check before you need the stuff.!

Normal air was 70% nitrogen, 20% oxygen and 0.0002% helium. Around the base of the waterfall was a warm pockets of steam. He had just proved one levels did massive amounts of natural molecular nuclear fusion. Totally nontoxic power! From regular water.

This was going to change everything! No expensive or oil or gas. The oil boys have plunged the world into recession of the 1970s. The gas moguls while working on the recession of the teens!

But there was no shortage of water around the world. There most of it has salt in. He laughed inwardly to himself-decelerating sea water was so simple.

You erected a buoyant tower that extended one metre above sea level. You had lower power pumping: A solar pump, or an animal tread mill would do just great. Though in the developed world mains power was probably the cheapest option.

The power requirement of a pump was equal to the pressure and volume of the pumped fluid. Who you were extracting massive amounts of pure water vapour at only 0.99 atmospheres. So he only needed a pressure head of 0.01 atmospheres.

You transported the water vapour to above a storage reservoir. You then vent IS THE gas to the air.
All the water condenses out! No salts or bacteria was transported in the water vapour. So you got effectively pure water.
You could add chlorine and bromine if you so desired! To protect against infections and pregnancy. The adding of bromine to tap water had surprisingly never caught on.
He did the obvious-well obvious to any academic who did not have a pressing need to fabricate an academic paper ready for publication. He had a cup of T. Having just proved that water it did molecular nuclear fusion, he wanted to check it. By boiling some water. And he immersing a T bag in the hot fluid.
He took the reading with a Geiger counter. He was so pleased the last forgot to stire the T before removing the back. He decided he should build a T making the machine to avoid this problem.
Academics doing distraction two are life threatening degree. He was fairly sure the Times crossword did not have such important theoretical threats!
Also he was thirsty; all this Olympic standard mental exercise deep rather risk the hydration. And a secondary use of all this the T was to quench a thirst! OK maybe that was the real reason. He just wished his own brain would not be such a pedant.
The last thing an academic needs is an unconstrained brain ranging lose about the laboratory. Even if it was their own brain. A very dangerous overture and

scientific instrument. All that was what he told his girlfriend to avoid doing the washing up!
Waterfalls warmed the air! And made the air electric. And produces helium gas. Obviously it was generating charged Alpha particles. Plus the extra oxygen atoms. They were not the quiet scenic area to avoid doing academic work. They were very scientifically interesting in their own regard. For too long scientist have regarded nuclear fusion as the province solely of physical experimentation.
But if nature did nuclear fusion then plant was the province of excrete every scientist alive today! And he is really mad moments that is what he considers himself to be.
He also has a looked for any excuse to avoid doing any work around the university. Ogoing cafeterias and free newspapers was an inducement. But to get these things you only needed to train as air crew. Though he had heard talk that people expected even pilots to fly a plane.
But these days flame was the province of the computer! When he was 14 he had programmed his ZX spectrum to fire a plane. And when only produces three catastrophic crashes that would have annihilated life on this planet.
He was working on the are now waiting life on Mars, until he realized that the radioactivity exposure all astronauts even prevented life astronauts making it to the Moon.
The NASA Apollo missions have been such an obvious Hollywood fake. The only reason that scientists had not

called 'a fraud', was that they're all wished to get some research funding from NASA.
Academics were not about truth! When never about truth. They were about where are the next piece of research funding was coming from. And that was what he was meant to be arranging now.
  He scrawled on his napkin:
  $2H_2O + T \rightarrow He^{2+} + H2O^{2-} + O. + E + \gamma$

Curious indeed! Nuclear fusion from regular water. By fluid turbulence. Just then Neil came over. 'are you making up this evening?' He regarded his laboratory partner. 'Hmm' was the reply. Neil had friends in physics, and this was looking suspiciously like nuclear fusion.
Which any idiot realized was the exclusive preserve of the physics department. The trouble they had had game biology to drop all interest in the production of helium gas by photosynthesis.
Physics had ended up directing biology to chemistry! And biology did not understand chemistry. And realised they were so out of their depth with anything regarding nuclear fusion. Physics could tell them that the start. Physics might not understand nuclear fusion: but at least they could draw up meaningless equations about Beta Masons, and scratch there by its own over agoing way. For this reason, growing appears was an essential characteristic of physics professor. Also to say 'Hmm' in a fashion which suggested they actually under stored some great truth about the universe.

Really they were just trying to solve that day's Times crossword. A PhD student has started doing them in 5 minutes, I had to be persuaded that is talents were desperately needed in Life Sciences.
'look at this!' Neil was commanded. 'what is it?' He enquired. 'Fluid turbulence' Jake replied. 'interesting' Neil observed. Ignorant about Chemical Engineering matters. He was fairly sure that fluid turbulence had very little to do that at pressure in the sports car. You're not allowed to concede total ignorance about any area of study and some other academics was doing.
'Fluid turbulence induces nuclear fusion' continued the chemical engineer-match to Neil's bemusement. 'But every sun in the universe a boiling hydrogen plasma does atomic nuclear fusion' Jake finished. Just as his T addled brain was proposing fluid turbulence of water also caused the nuclear fusion of hydrogen atoms: bound up in the water molecules.
Though it had to be conceded PG Tips were the experts here. Though there a problem totally unaware that they were the cleverest people on the planet.
'look I have got the first year practicals to supervise' Jake suddenly realised. 'Let me work on this' Neil showed to him as he left the room for a frame in a tornado year of chalk and T dust.
Jake diverted via his office, in two minds whether to make it in the practicals laboratory the day-or just give the students a free afternoon.
Actually do some work! Then he remembered. Academics don't use it to work! They have developed a

section and elevation and incompetence to mind numbing levels.

Undergraduate particles are the ideal time to read the newspaper and do the required crossword. And catch up on your letter reading and writing-and obviously make up your sleep hours for the week.

In the evening all the drinking there were required to dear tended to erode the answer available for sleeping. The biologists have that one sussed. They ran practicals and slip technology. And got paid to do 40 winks.

All in the name of scientific progress. And nothing whatsoever to do the universe C drinking semi final at was looming eminently! The university has a world wide reputation for the departmental drinking groups.

Which was a very nice way, to say a lot of drunks worked here. Hence the major headquarters of AA by the sleep laboratories.

They are obviously work had to be applied in its most elastic connotation. To cover the hours of sleeping a day they had to do.

It was lucky they are so much undergraduate marking to do. Academics has realized for decades that certain activity fdo not usually require you to be awake.

Not for nothing do the university have such a reputation for somulence and drinking. His undergraduate career, is so not being obvious when he was destined to be a world championship sleeper or drunkard.

He never considered there was a contest! So it just became both. Only having to sober up, and wait for an action for a or as during term time.

Biology

Against his better judgement Jake was doing some marking. His least favourite task. The brightest students would insist on plagiarising Internet papers. Luckily he has read all the relevant papers, so could mark down the undergraduate essays and write 'don't plagiarise' in red ink.

If any student ever came up with a creative idea, he you plagiarised it and gave them a C+. He was still researching his back of such essays from his undergraduate years.

And family all that drinking have paid off! Yes, some incisive creative thinking. He recalled at the time, he was struggling with the assay as they just had a A Jak with the paid for marker.

He was not being clever, he was being humourous. He mentally resolved to take his sense of humour into a padded chamber and give it a good kicking! Creative thought was not allowed in higher education.

In primary and secondary education that sort of thing was usually went off with only a police warning. Needless to say, he was the only primary schoolchild he knew of with three suspended sentences. Mostly the split infinitives in.

'Only plagiarise when you got to be a university lecturer' was the best free advice he ever gave his students, when there were so work and he learnt enough to understand English. When you are a member of the teaching staff ; you will have given up all hope of doing original thinking for yourself. So borrow the mind of your

undergraduates. Far less wear and tear on your own brain.
Jake had always had to try and remember, one day he might need his brain. Though usually not at university. That clever Ph.D. student had push you on solving the Times cryptic crossword for almost a week.
Nobody was allowed to be more proficient at the cross eyed than herself. He landed after only two months, the troublesome Ph.D. student was a nephew of the compiler. That was allowed -teaching is just using your brain!
Except for the occasional necessity to do original thinking , Jake liked being a biology/chemist! He rather liked one particular lady biology lecturer. He could get on quietly with his marking, safe in the knowledge that though pretty, she would not say anything of importance! He has spent the last six months trying to her and had his wit, so her only treat all would be 'your bed online?'. Obviously the undergraduate was not completely dead and in yet.
The last important thing any biology teacher has said, was that plants and animals are the teacher! And were in a competition for survival. Rather like kids in a primary school! Only primary schoolchildren are usually less violent.
He went to sit in on the end of the leading biologists - Lynn's, first year undergraduates biology lecturer. As a tea shop was having an explosive be painting, and any lecturer and sat down will produce and so light on the wall, and get covered in paint.

They were doing the common room cyan, and he had always preferred A yellow and blue pokerdot-with psychedelic extensions and pretensions focused on reality-Jake try it to do that at least every other year!.
'so green plants take in light produce; helium and emit gamma wave radiation' Lynn told the semi somnolent undergraduates. Jake nearly died! His hand shot up. OK, this was breaking etiquette but this was important chemistry and biology!

At job it wasn't physics, of Jake do not do physics. But the Physics had never quite got over photosynthesis giving off nuclear radiation.

Any chemist would have happily told them, is no chemical source on nuclear radiation. Only nuclear fission or fusion can give off nuclear radiation. And with no radioactive element, only eight nuclear fusion from compounds of hydrogen was in the frame.

There is an unwritten rule of academia, you don't comment on the validity of science taught by other disciplines. So medieval Latin rarely commented upon music recitals. And everybody had a good laugh at physics!

The but then physics got $100 billion to run and build the great hadron collider. Whose only achievement is to highlight a boson particles, that was a massively two light and far too rare to give all matter mass.

In the 1920s Einstein had shown that mass and gravity were another dimension of reality! The gradient of the gravity slope gave you the local gravity field. And matter reacted with gravity to produce mass.

So why was physics looking for particle which endowment mass to matter? Who was stupid politicians would give them a blank check to build Esoteric Equipment.
And demonstrate nothing of importance. Since the 18th century when Newton plagiarised another mathematicians work, the only original thinking have belonged to a Swiss patent clark-called Albert Einstein. He have been wrong in any direction. He ievery the speed of time to make object masses of forces to line up. He should have buried the permeate energy of free space.
This determined the local speed of actions at any point in the universe. Gravity slowed down our local time, from the university time constant. So gravity slow down the passage of time-universal time remained resolutely constant.
And that was the most incisive fault in physics for 200 years? I think that is all we need to say about physics. Jake stood up and pontificated 'sorry to protest Lynn, but the helium and gamma wave radiation are they exclusive province of nuclear reactions' Jake explained as if to a secondary school idiots.
This he had been taught in A level physics. He has had a real problem explain to his undergraduates, that biology understored carbon dioxide! Physics and engineering didn't. It was just how they got money from nuclear power. Professors of physics road fictitious papers about carbon dioxide-as if they actually under stored for the are talking about.

Lynn had suffered enough under A level physics, to have a sli to understanding fact helium and gamma wave radiation were the province of physics-not biology.. She rocked back on my heels, and deleted the last three sentences off the blackboard.

Reality did not usually intrude into academia. Where academics have free rein to talk about pretty much what they might , and he cites she had three warehouse of chalk to use up.

If Jake wanted her lectures to be real and true for all. There were 20 lectures are fears that she had a arguments with. All the students did, when there has so we had enough to string three coherent sentences together.

Only when they had used up that chalk allowance could the university teaching staff join the 21th century. But only if lectures in other disciplines did not sit in the on her lectures. What was Jack's problem? She had thought he rather fancies her. He never warned her that even actually listen to what she is saying.

Lecturers could say what they wanted! And undergraduates were manmade sli rewrite their faults in exams. Information passing from out to written words that passing through the brains of ur the drunken students.

'could be any body explain the question?' Lynn frantically asked the students. As she lacked the instant answer you expect your lectures to have. This was why disciplines get the hell out of other sciences areas. Because they really knew what they're talking about in their own discipline!

'Jake please explain!' She frantically asked the questioner to answer his own question. As she could not. Why should she be doing all the work? He was just doing the crossword and his marking. She would ask about five across later.
'helium gas and gamma wave radiation are the exclusive province of nuclear reactions. So you must have a source on nuclear fission around' he retorted. More than a little annoyed to be asked to answer his own valid question. His a level physics teacher had at least prepared him to say something vaguely reasonable on the subject.
'life abhors toxic nuclear fission. So we can guarantee that the production of helium and nuclear radiation around the world is not linked to nuclear fission' Lynn protested. Wishing that she had stayed awake and not been so interested in boys during physics.
She has spent her of physics lessons turning up on the finer points of sexual reproduction. This after all was biology-and that was going to be heard degree. Or boys. She had yet to decide. Only at university did you realise the two areas were not mutually exclusive.
'we also see the production of ozone' she added. 'there are chemical sources of that gas-mostly photosynthesis. Which does produce helium and get our gamma waves' she conceded.
That was written up in the postgraduate documents on photosynthesis. There never explained. As for biologists did not have a working knowledge of physics. 'did sir hear that photosynthesis only gives out gamma waves, helium and oxygen in the light' the university

cheer leader added. What use was are being drop dead gorgeous if a sex bomb could not ulcer think.
'molecular oxygen, and free radical oxygen?' Jake asked the 19 year, Anderson look-alike reject. Molecular oxygen was O_2, free radical was O. Helium reacted with nothing. It should not exist outside suns. Outside regions where nuclear fusion was going on.
'I can answer that one' interjected a marginalised Dr. Lynn Fornight. Who was feeling that the lecture was getting away from her. 'the O results in the formation of ozone only during the light' she informed her students and herself, as the bell sounded.
The students were rushed out for lunch-so much more important they thought than simple photosynthesis. Which does supported all life on earth, and ensured any additional carbon dioxide from the fossilised life at the end of the Jurassic, ended up as additional life in the modern environment.
The fossil fuels were fossilised life from late Jurassic: an undergraduate final year thesis has concerned the fossil life found in the fossil fuels: the shells of fossils were exclusively from biology which had died in late Jurassic. They can be positive about that from London stratum with the same fossils in: just below the worldwide soot layer that showed where the comet in Mexico had set of world wide forest fires.
Ever since grass had encroached on the forests, by being the most farm of all biology on earth! They emission of gamma wave radiation during the light do is a summer, ensured there were regular buyers across grassland.

This devastated by certain trees, and a short the new biology which regrew was based around grass, as the major plant life. Trees were for ever marginalised! They were the most pernicious life to have survived from the Jurassic age. When there were a world wide forest covering even the land based South Oole.
No increase of the trace of carbon dioxide in the air was biologically possible. All through history the ice cores and mineral records show that free carbon dioxide levels can only increase during a natural ice age. When there was less photosynthesis.
'so' Jake wrote on the blackboard:
$$CO_2+3H_2O->CH_4+He+E+\gamma$$
He had the forlorn hope that biology would pay him for his chemistry. This was always likely as the sun rise in the arctic documents. Never going to happen!
Okay, it had taken him 10 minutes and eight alterations to devise the equation. Lynn had left for the cafeteria. After all she had to eat, and have lunch was waiting in the senior common room. It was Lasagne day-her favourite!
Jake was two preoccupied with his stroke of genius, to note is a steady trickle of students leaving to eat at the junior common room! A less salubrious establishment, which played Oasis and Blur rather than Beethoven and Baché-younger and older!
So biology does nuclear fusion from water, catalyzed by carbon dioxide! That imposes a total limit on global average carbon dioxide levels. Nature has been doing molecular nuclear fusion from water for 3.8 billion years!

Physics is a fusion torouses I have failed to do exothermic nuclear fusion ever.

They had just introduced a perpendicular magnetic fields to give the hydrogen ions spin as there is circulating around the torous, this helical flow would have induced atomic nuclear fusion.

Physics have a fault of this on in the 1950s. But I kept it quiet, to maximize their monetary income from nuclear power to research rubbish about the climate! Nuclear power has apparently never noticed physics do not understand organic carbon or the weather! A win win or research area.

He announced to the NT lecture room. Looking round he cursed, jotted the insight down on the reverse side of his marking, and rushed to lasagne day!

Apparently hunger is more important to undergraduates them gram breaking science. They do not understand the science, but have had many years experience of hunger!

Susan trudged heavily back to her digs: singing happily to match her mind frame. Jake had given her stuff to think about! So she thought about that, if she was really bored she might even consider nuclear fusion!

She was staggered: she couldn't believe that physics said nuclear fusion went on in all life forms! She had done a Google on it. All biology produced helium, free radical oxygen, gamma wave radiation and energy! By the turbulent flow of water in plant photoblasts. Or the turbulent flow of blood around the body. Really just around turbulence!

Or it could be induced by carbon dioxide metabolism within cell mitochondria. As an undergraduate she had been told that we breathe out helium and methane! This seemed so far out, she discarded it! After all there were no compounds of helium accessible to anybody! Other than Ph.D. chemists with access to high pressure for ring and helium gases. So nobody that counted.

The presence of helium within biology was two impossible! Even a university lecturers would never contemplate biology could do nuclear fusion. Unless it is heated to 10 million° C!

She stopped at stinks! The biology department. Undergraduates has simplistic mains for the departmental buildings. Chemistry was Bangs! Biology, and its organic refuse recycling power plant, was called mega-stinks by even the general population.

And the sociology block was called 'exit to main road'. For obvious reasons. People needed to get the hell through that building as quickly as possible!

This socilogy block have been cleverly designed, so other student studying other subjects need never know it even existed! It did contribute income to UMAT, and was a perpetual source of free university degrees if the toilet paper ever run out.

'Hi Tina' says Susan as she recognise her friend. Well maybe friend is a big word for a little person! But calling her a tame midget was turned politically uncorrect these days. The phrase 'political correctness' was coined by the Klu Kluk Klan, and frankly, they will welcome to it.

All of those trendy lefty political students, is in the phraseology invented by black heating Hicks from rural American-pretty much the same sort of individuals!
Dr. Tine Fookes was their head sthink for the organic refuse recycling centre. This masterly undercuts the local water company, by using hard science and under water aquafer, so they had ended up supplying water to the three nearest towns and cities. Manchester, Salford and Bolton! But frankly, Bolton had got a bit lost on their geography. The largest town in the country! Never made a city, as other conurbations hated it. And since they got rid of big Sam, the football team has not been so great either.
'have you heard about Jake's ideas? She asked Tina. Ms Fookes have never changed her surname, because of family were very proud of their Saxon name. And took 'what the fuck' as a compliment.
'he is a chemist isn't he?' Asked Tina. 'that's right-you have heard of him!' replied Susan. 'not really-but I read the recent article in the university magazine. "ecological science is just a pile of crap"' she repeated locally.
She needn't have bothered! The whole university has made an agreed with the article. It was definitely going to make jake Mr. unpopular for the rest of the term. Why then they were going to explosions at recycling plant, and everybody would have become distracted by the sudden, though financially lucrative undergraduate deaths. F
'he was misquoted' she defended the guy. There she was sure the chemistry department employed people to

defend their stuff, using short range tactical nuclear weapons.
'when you see them, you can tell him the sewage factory has only had five explosions this week! We may be after the 'Bang' department name' the sewage girl threatened.
'he reckons photosynthesis takes in carbon dioxide' Susan expanded. 'as taught to every high school pupil in the world' notice Tina in a bored mindset.
'he says that organic carbon is the limit to life on earth!' Susan explained, waiting for some biological correction here.
'but the physics guys tell us photosynthesis has "topped out"' Tina mindlessly recited. It wasn't true, but what ever brought money into university she was not about to argue with.
It doesn't matter how inaccurate ideas are, just as long as they bring money in to UMAT and help paid the wages. Biology departments the world over have been trying to help physics with the carbon cycle for 30 years. Physics have gone selectively death. And refuse to acknowledge that global photosynthesis imposed a total cap of free carbon dioxide in the air. Biology was the older science, and felt so marginalised.
'next you're going to tell me there is no external source of carbon' she factor asleep enquired. Knowing damn well that all fossil fuels used to be active biology in prehistory. Though obviously history was not her major research area.
'all the fossil fuels formed at the deaf the dinosaurs 65,000,000 years ago air' Susan proudly told the biology

lecturer, who apparently did not know it all! There lectures usually boast that they do.
'there were none formed in the cretaceous of Permian? Tina checked. 'during the 1000 year ice age? Very little active biology' Susan declared.
'all the fossils in the deep coal or oil reserves around the world, are exclusively from biology that they had at the end of the Jurassic age' she reminded her. Certain that any decent biologist would know this.
'we have been trying to tell physics that since 1986: Chernobyl took precedence over scientific fact. And has done ever since. Can she go and tell physics that' pleaded Tina.
'he said selective deafness? More like permanent deafness do anything which endangers the research income stream from nuclear power' Susan finished.
'so what science has Jake come up with? I am so bored about clubbing about carbon dioxide-the gas of life!' Tina exclaimed in exhaustion.
'you know that life gives out helium and gamma wave radiation?' Susan started expanding on the idea. 'an methane, and the free radical oxygen which results in ozone formation throughout biology' filled in a world weary Tina.
'what you may not have realized, is that helium and gamma wave radiation are the end results of nuclear processes' elaborated Susan. Tina was so shocked, she almost woke up!
'but there is no toxic nuclear fission throughout nature-all around the world! There is only one small rodent that

utilises regular uranium in it's auditory canal' protested Ms Fooke.
'we aren't talking about toxic nuclear fission! This process goes on in your cell mitochondria, and your arteries and in your beating heart' she expanded.
'not nuclear fission?' Checked Tina.
'nuclear fusion-from regular water' explained Susan; and was pleased to see teen years had your from the open in amazement. Either that or her waters had just broken. She didn't look nine months pregnant though!
'the turbulent flow of high pressure water or steam converts H_2O into helium and oxygen gases, loads of heat and a little gamma wave radiation' she explained.
'so that is a source of are on bended oxygen' noted Tina.
Here tea break would have to wait. This was locking important! 'so the methane gas we see throughout nature' enquired Tina.
'their results of nature metabolising carbon dioxide and water, to form methane, helium and oxygen' Susan finished. Feeling quite pleased with herself, that should remember it all out chemistry rubbish that Jake had enthused about.
'no 10million° C' Tina asked, like she actually understood the idea. Really a light bulb was going on in their brain, only slowly-and she followed her brain with off peak electricity from the mains.
'down to 3° C-with arctic fresh!' Susan informed Tina 'that is my research area' she filled in. 'show me' she commanded Susan. Susan sighed. 'you could just ask Jake' she reasoned. But Tina wanted more!

$$CO_2 + 3H_2O \rightarrow He + CH_4 + 3O$$
'So methane, helium and oxygen' Tina nodded. 'ground breaking stuff! When does he publish?' Tina naturally enough replied.

'I guess that depends on the engineering and physics department' Susan replied. 'why not biology-the a Ph.D. is breaking do on this area! The money from global warming is know nothing compared to the funding we can get for nuclear fusion work!' She reasoned. Susan nodded.

'are you bringing the quiche to the women's group tonight?' She asked Tina! 'oh yes, I am! Thanks for the reminder' the sewage girl finished.

The right then, the physics department was faring significantly less fragrant than the biology of state for us it. As to explosions-watch this space!

Biology is over 300 years orld is a science. Physics is 250. But professors of physics always claim nuclear fusion as their own. Despite not having done exothermic nuclear fusion in last 50 years and $500 billion spent on failed fusion torouses.

A professor of physics need only take a walk into the countryside with a Geiger counter, to see that green plants do nuclear fusion down to 3° C or even lower! All for free. No massive failed fusion torouses.

Physics

It is a tragedy that physics do not leave until two engineering! Any engineer around the world could have expressed the importance of fluid turbulence inductee nuclear fusion.

So lands around the universe all have a gas plasma in turbulent flow! Initially hydrogen, but as a fusion from grass is latterly helium. When the gas plasma gets to lithium, the sun ceases to shine. It contracts and whizzes around the universe as a dense plantetoid. When the gas plasma fuses all the elements are to iron, or nuclear fusion stops! Your together two plutonium and uranium before nuclear fission gets to be exothermic.

Biologists could are taught physics professors, that the animal hearts and arteries give out nuclear radiation as the heartbeats. And the are more breeze out helium gas, though the only breaths in 5.125 parts per million in city centres. He also frees out methane and free radical oxygen!

So biology is knew stuff, but did not know what they knew. Helium gas and gamma wave radiation are only a produced by nuclear reactions. And there is no toxic nuclear fission going on in the body! Certainly none in time with the heartbeat!

Animal blood systems do nuclear fusion! But the physics never talked to biology. It would just take in one phone call. But academics pride prevented it.

'that explains so much! Like the emission of radiation from boiling water' Neil told Jake as he re-emerged in need of a biscuit and cup of T.

'how did you come up with it?' Neil asked him. 'well I went off to the peak district with my Geiger counter, to see if there were any radioactive rocks that might make a good practical' the engineering lecturer lied.

'good man' Neil complimented him 'the most insightful work of technology ever really' here enfused.
'I looked up the strong atomic force for hydrogen' Neil carried on-being the good physicist he one day he hoped that she will be. 'in molecular form?' Jake checked. 'Ah, anyway'
Neil carried on.
'a one metre drop will cause some of the water molecules to overcome the strong atomic force for hydrogen'. He preferred a statistical diagram showing how the turbulent flow overcame the atomic force for some hydrogen atoms.
'it is the top all the bell curve, for all the interactions of the water molecules'. Jacques looked puzzled.
Neil explained how they chaotic interaction overcame the strong atomic force, which prevents water in a cup of T from becoming helium gas and dried T. 'this is so much more powerful than the interactions for hydrogen gas' he finished.
'so that the linear flow of hydrogen gas was never going to give us atomic nuclear fusion?' Jake checked-trying to be engaged the conversation.
'statistically only a proportion of the water molecules will undergo molecular nuclear fusion. But the interaction is 20 times as powerful as the hydrogen atoms in turbulent flow air' Neil finished.
So all the money spent on fusion torouses was a waste of money. All they needed to do was to introduce spin on the hydrogen ions go around a small fusion torous. The major fusion catalysed on suns was the chaotic flow of the boiling hydrogen and helium gas. Just as his flask

of boiling water high give off nuclear radiation as the water boils!
Jake wandered around Sheffield with his trusty Geiger counter, checking on boiler rooms around the city. He always taught the boiler rooms were not safe places to hang out. Now he knew why!
Boiling water gives off gamma wave radiation-and produces an in fall totally small amount of helium gas. As a gas analysis at Sheffield University demonstrated. They were the world leader in such technology. But nobody had thought to apply it to boiler rooms before.
Dr. Turner came into the room. 'ah the very man' he exclaimed. 'am I?' Jake replied. 'probably. The uncertainty principle means you can never be totally sure in this universe! Cecil told him.
'what can I do for you?' he asked. Filling in at practicals he could about manage. After all engineering was just practical physics! And physics was philosophical thinking, or drinking cups of tea.
'this molecular nuclear fusion stuff of yours' Cecil started. Growing herself up to his full height of nearly 5"3'. 'it will never do! You must stop it' Cecil told him in no uncertain terms.
'why?' quaked Jake. 'bally climate has out of age or' Cecil informed him. 'never do. Don't want to change the universe-only by seconds of arc' the Dr. of physics told him.
'but I thought you wanted to change stuff!' Jake protested. 'then we'd be out of a job by next year' Cecil informed him. 'only change stuff slowly'

'that global warming stuff was brilliant. Couldn't possibly be true: all the historical data showed no causative link between organic carbon and the climate. While the plebs for carbon mattered to the weather we made a fortune'
'but the weather cooled from 1997!' Jake protested.
'yes, damn inconvenient of nature at one!' Cecil reflected 'but then we had climate change'
'which said?' Jake checked, as he had never quite been on to figure out what it all meant. 'quite obvious really! Carbon dioxide causes the natural weather' Cecil lectured to him.
'like the natural weather before mankind evolved?' Jake asked. 'yes that's the one. I only know totally caused by man. And burning carbon fuels' Cecil continued, less convinced by the minute.
'so in other words, carbon has no effect at all!' Jake thought it through. 'Shush, don't want everybody knowing. After all we are talking research grants here' Cecil desperately observed.
'and then you go and give everyone nuclear fusion. Looked play their game' Cecil demanded. 'which is?' Jake not unreasonably enquired. 'agree with any old rubbish providing we get research grants!'
'so global warming was never true?' Jake checked. 'did you never do O level biology? Biology experiments to use up all the available organic carbon. Any local shortage are minerals does not matter to nature' Cecil wearlly told him.
'and then you go and give them nuclear fusion from water' Cecil continued. Jake was taken aback, the

university Grapevine obviously worked better than he knew.

'the computing department has monitored your googol searches. They routinely have to vet all Internet posts. Molecular nuclear fusion-great idea. Just not allowed' Jake was told.

'if you persist in this research, your university tenure will have to be terminated.' Cecil heavily instructed the engineering lecturer.

'besides you have it all wrong! Or maybe not wrong. You been doing a lot better. I will have a word on your behalf with the heretics. About getting you into the nuclear fusion sessions' Cecil told him; informing him for the first time that There was such a group .

'All I can tell youUsed to read her on gas plasmas. Particularly hydrogen and helium plasmas .While you are at it read up on lightening.'The old man said as he left the room.

Environment

Susan hated the word strutted! It was a sign of academic over confidence. But since she had heard about Jakes work she had begun to strut. The only problem was, he was a chemical engineer. And they are don't tend to talk to Life Sciences.

In the same way that biologists and physicists Rayleigh share a golf team! They don't talk. They are in competition for research funds. They were both trying to save life on earth, so what the hell was their problem? She walked carefully, slowly and did not strut into the pink concrete eighties monolith. If Jake was not prepared to talk to Life Sciences, then it fell to her! She quite liked this, as you would get all the plaudits! The animosity between the two departments was an incurable. Whatever.

'you are a short this guy isn't the antichrist Susan?' asked Dave spades.

'I know he is interview in the magazine was a bit of an own goal, but he does saying interesting stuff' Susan pleaded. 'OK-but he can expect an argumentative lecture' Dave warned.

Jake finally found the Life Sciences room. He has recently penned a scathing letter about environmental sciences being about as much science and gold fish and scientology lectures.

'ah Dr., We are all ready for you. Unfortunately the Lions are taking the day off, otherwise we would obviously have thrown you to them'. Welcome him in.

'you read my letter then?' Jake apologised. 'yes. So I was planning a combative lecture were I plate the bad cop, and you the helpless accused' Dave warned him. He smiled at Susan. Let battle commence.

'the most important fact for you 'scientists" he got out without collapsing in laughter 'to appreciate is that nature is full of helium gas production and the emission of gamma wave radiation and visible light.

'just to stop you Jes, what is the importance here of visible light: though obviously as card carrying scientists all the undergraduate scientist in the room already know. That remind us' Dave instructed in only a moderately hostile fashion.

'A very important interruption David. Thank you.

Chemical processes do emit electro magnetic radiation up to the lower infrared' objece explained. Fairly sure nobody had ever called him Jes before.
'and that is' David demanded. 'low infra red warms stuff. He can't see it, and you would struggle even to the iit' Jake expanded.
'if I could stop here for a few minutes there Jes; I did a world recognized survey of light emission on by a bacteria in the deep. Though I am sure all real scientists hammer out and assimilated that paper' days proudly told the hapless theJake.
'I will return to that point later. Neither the bacteria not chemical processes and in the deep could generate visible light if there were not doing molecular nuclear fusion from water.' he countered, thinking mentally that letter to the union magazine was not the best idea he had never had!
'go on.' Instructed Dave, smiling and thinking he was going to enjoy us. Vengeance belonged to Life Sciences. As he was sure Jake would soon realise.
'all through nature we have the turbulent flow of high pressure water or steam. Like at a water fall, deep sea currents or breaking waves.
The water molecules are in chaotic interaction, and some of the hydrogen on hydrogen interactions exceed the strong atomic force' Jake lectured. 'wow there-a physics free zone here' Dave reminded him.
Jake nodded. He had been expecting a question from the floor at this juncture. Dave was obviously fillling in for the puzzled undergraduates. 'the strong atomic force stops atoms from fusing. Otherwise all appreciation

would collapse back into the infinitely dense state before creation' Jake explained. Omitting his doubts as to whether god had ever arranged a big bang.

Science had single he failed to explain where the infinitely does protein is has arisen from. Stephen Hawkins had expanded the idea that the universe vibrated to and from the big bang particle all the time. With no corroborating data, he had withdrawn the idea 25 years ago air. Still the universe was here. And have been for 14,000,000,000 years.

I was based telescopes looked further into space, that big bang instant was pushed back five billion years every 4 years.

'all the time nature generates helium gas' Jake trotted out, sure of the science here.

'how do we know that the helium gas does not just loiter around? And how much is there?' The David interrogated him. Warming to the task of making this chemical engineer understand there were other sciences out there that he did not know about. Jake had always thought that Life Sciences were ejected. He wasn't all that opinion now!

'hydrogen and helium are not held by the earth's feeble gravity. Lightning reacts hydrogen with oxygen to form water! All the time around the earth.

But helium is lost to space. I have gone on record before saying there is 5.125 parts per million in the air.' Jake trotted out his established science.

'hold on! That is over twice the average carbon dioxide in the global air!' Noted the life scientist. '2.5 times the level of carbon dioxide Indies' Jake confirmed. Carbon

dioxide is metabolised by photosynthesis. Producing helium while emitting a broad spectrum of electro magnetic radiation' Jake told the bemused undergraduates.
'life science has known since the Chernobyl incident, that photosynthesis exerted a total limit of free carbon dioxide in the air' Dave confired. 'why did you not tell us?' Demanded the chemical engineer.
'I thought all real scientists on earth knew that!' Countered David. 'we just forgot we knew that' acknowledged Jake.
'all around and even in us, nature is turning water into helium and oxygen gases-with a massive release of heat and nuclear radiation' finished Jake.
'including visible light and radio frequencies' checked Dave. 'indeed. That is why it is important we gain hydrogen from space, so at lightning strikes we get back our water! To go around again' Jake told him;
'so this system operates wherever there is high pressure water' David not unreasonably asked him. This was ground changing stuff, and a you wanted to ensure that Dr. Jake the Rate had got his facts correct.
' your own blood stream does it.' Jake instructed him. 'not just the cell mitochondria-I have known for years they give off nuclear radiation. Though I did not appreciate the importance of that fact. And physics showed no interest' Dave lamented.
'your cell mitochondria do and do molecular nuclear fusion, utilising the carbon dioxide in your own blood. But the pressure waves created by a beating heart also MNF'.

David looked at his watch. 'Sorry' he apologised 'we have run 5 minutes into your free periods. We will pick this up on Tuesday' he dismissed the undergraduates. 'so your contention is that all biology and nature do nuclear fusion from molecular hydrogen?' He passed the chemist, warming up to the skies. Yes make he usually burnt at the stake-but after they had extracted the information from his brain!
'that's the idea' Jake conceded, failing less marginalised-maybe he had won Dr. Kurcher over.
'not at 10million° C?' Checked Dave. 'Down to 3° C-the maximum density of water. It expands as it turns into an ice' he reminded David. 'yes yes yes! There are real scientist in this department. We have been trying to remind physics that an ice melts it shrinks for 30 years! But all we've got for our trouble is some for this" he complaint. 'this is world changing science. Want to write a paper for Nature?' He stunned Jake by suggesting the right science together.
'after I get back from addressing SERC that would be great' he happily admitted.
In his lonely room Jake was gazing lovingly a lot up and up to the notice board. "you are cordially invited to address the science and engineering research council" it shouted. Well maybe it whispered it, but in his mind it shouted.
He have been working on his address for last four weeks. Even got in the way of this practice for last bridge tournament. Which is why then only, 28th, at of a field of 26.

This was going to be the most remarkable so the state of his life. Hopefully also are the most remarkable science day for the world!

SERC

'... and so I can tell you, that the kinetic interaction of water molecules does molecular nuclear fusion.' Sheikh finished off his address to the assembled scientists.
'you are talking about heavy water! Duterium and tritium enriched presumably' Dr. Filamore asked him. The good Dr. had secured six million euros to research heavy water.

What Jakes have proposed was about to end all nuclear fusion research! By actually giving the world access to the nuclear fusion which are already whitter on around the world. Even in their own bodies.

Needless to say, the fictitious climate research would end of the careers of many individuals. This strictly speaking by nuclear stooges, and not scientists! And yet the world's allowed them to lecture the greatest brains on the planet with nuclear fiction.

At least the good Dr. Was not shooting total Blanks! He was already thinking of evaluating the heavy water production of more molecular nuclear fusion cell. Bubbling steam through high pressure water should do measurable quantities of molecular nuclear fusion. Is it is pretty sure already that bubbling steam through water gave off nuclear radiation. Though the quantities of helium by too minute to have measured.

Unless you knew what to look for! And physicists have become a very adept at only seeing effects which would result in large research budgets. They didn't like nuclear power-just too dangerous and toxic. But they were the only individuals were blank check books paying physicists to research and the.

Lionel Filamore was certain that no democracy would never allow the construction of a nuclear plant ever again. Things were just recovering after the Chernobyl incident. Then the world did Fukushima.

Physicists had never been a water may get the case that the Japanese were in some way technically defective. The Germans and Japanese were the most technically proficient nations on earth. Which is why they came second in the last world war.

Mind you, taking research money from nuclear power was more dangerous activity than trying to juggle with red hot devils. Not a sport he then it don't you understand. It was just his regular nightmare whenever he wanted to avoid coming to work!

'no heavy water. Just your regular tap water around the world' Jake corrected him. 'though we need high pressure water or steam with one atmosphere turbulence to start nuclear fusion' he gloated. After all this was the most important science of his working career. He scrawled ^{1}H on the blackboard to reinforce his point. Just regular helium. Not be $^{2/3}H$ in heavy water.

'so power with no carbon dioxide?' The Dr. checked. 'global photosynthesis reverts all additional carbon dioxide to the active biology in the active life from the Jurassic age. There is no external source of carbon-it all used to be active biology' Jake corrected him.

If physicist could have embraced that basic science in 1986, the world would never have heard of global warming. Global photosynthesis converted all additional carbon back into active biology it was in prehistory. But then they could not have got massive research budgets- looking into are totally spurious science. They told the world was life important science. It was really fiction on behalf of nuclear power.

For too long the supposedly greatest minds on the planet had totally omitted to factor global photosynthesis into their research! Favouring instead the latest fiction from nuclear power. Science is not about truth. It is amount obtaining the largest research budgets.

'so you are giving the world coldfusion' Fiona asked the engineer. He winced at the very mention of this discredited science.

'room temperature nuclear fusion other compounds of hydrogen we have one atmosphere of turbulence in the fluid' Jake corrected her. 'Room temperature nuclear fusion' would have been a more snappy title.

'when are you going to publish' Fiona past him. 'I would be interested in drafting the paper for nature' offered the Dr.. The climate change ship has so obviously saled. Time for new science.

'the world started using MNF' he used the abbreviation with no explanation 'at the advent of the industrial revolution' Jake announced.

'only the expansive steam cycle presumably' asked Lionel Filamore. 'obviously' replied Jake, not having realized until that instant this was probably true.

'so Chemical Engineering has changed the world. Into a better, safer, nontoxic, far cheaper sustainable world' Jake announced-only wishing the world's media have been here to record this instant.
His moment of defining genius!

Jonathan Thomason   JonThm9@aol.com

www.ingramcontent.com/pod-product-compliance
Lightning Source LLC
Chambersburg PA
CBHW070432180526
45158CB00017B/983